WEIRD ANIMALS
PANGOLIN

AMY CULLIFORD

A Crabtree Roots Book

Crabtree Publishing
crabtreebooks.com

School-to-Home Support for Caregivers and Teachers

This book helps children grow by letting them practice reading. Here are a few guiding questions to help the reader with building his or her comprehension skills. Possible answers appear here in red.

Before Reading:
- What do I think this book is about?
  - *I think this book is about a weird animal called a pangolin.*
  - *I think this book is about why pangolins have scales on their bodies.*

- What do I want to learn about this topic?
  - *I want to learn about what pangolins eat.*
  - *I want to learn about where pangolins live.*

During Reading:
- I wonder why...
  - *I wonder why pangolins are covered with big scales.*
  - *I wonder why pangolins have such long tails.*

- What have I learned so far?
  - *I have learned that pangolins often live in big forests.*
  - *I have learned that some pangolins are big and some are little.*

After Reading:
- What details did I learn about this topic?
  - *I have learned that pangolins use their claws to find food.*
  - *I have learned that pangolins have long tongues.*

- Read the book again and look for the vocabulary words.
  - *I see the word **scales** on page 8 and the word **claws** on page 10. The other vocabulary words are found on page 14.*

Look! I see a **pangolin**!

Pangolins often live in big **forests**.

Some pangolins are big and some are little.

7

All pangolins have **scales**.

Some pangolins use their long **claws** to find food.

All pangolins have long **tongues**! Slurp!

Word List
Sight Words

a	find	little	see
all	food	live	some
and	have	long	their
are	I	look	to
big	in	often	use

Words to Know

claws

forests

pangolin

scales

tongues

38 Words

Look! I see a **pangolin**!

Pangolins often live in big **forests**.

Some pangolins are big and some are little.

All pangolins have **scales**.

Some pangolins use their long **claws** to find food.

All pangolins have long **tongues**! Slurp!

Written by: Amy Culliford
Designed by: Rhea Wallace
Series Development: James Earley
Proofreader: Melissa Boyce
Educational Consultant: Marie Lemke M.Ed.

Photographs:
Shutterstock: Artem Avetisyan: cover, p. 1; David Pineda Svenske: p. 3; Stephen Bidouze: p. 5; Positive snapshot: p. 7a; Vickey Chauhan: p. 7b; Robin Bruyns: p. 9; Michael Lange: p. 11; K Hanley CHDPhoto: p. 12-13

Crabtree Publishing

crabtreebooks.com   800-387-7650
Copyright © 2024 Crabtree Publishing
All rights reserved. No part of this publication may be reproduced, stored in a retrieval system or be transmitted in any form or by any means, electronic, mechanical, photocopying, recording, or otherwise, without the prior written permission of Crabtree Publishing. In Canada: We acknowledge the financial support of the Government of Canada through the Canada Book Fund for our publishing activities.

Printed in the U.S.A./072023/CG20230214

Published in Canada
Crabtree Publishing
616 Welland Ave.
St. Catharines, Ontario
L2M 5V6

Published in the United States
Crabtree Publishing
347 Fifth Ave
Suite 1402-145
New York, NY 10016

Library and Archives Canada Cataloguing in Publication
Available at Library and Archives Canada

Library of Congress Cataloging-in-Publication Data
Available at the Library of Congress

Hardcover: 978-1-0398-0982-6
Paperback: 978-1-0398-1035-8
Ebook (pdf): 978-1-0398-1141-6
Epub: 978-1-0398-1088-4